T0063835

A RAINFOREST ECOSYSTEM
LEAF
CANOPY
TOUCAN
SLOTH
UNDER-STORY
SECONDARY CONSUMERS JAGUAR
SPIDER MONKEY
MANGO
MACAW
HELICONIA PLANT
HUMMINGBIRD
PRIMARY CONSUMER TAPIR
PRIMARY PRODUCERS FOLIAGE
SHRUB LAYER
LEAF
FOREST FLOOR
BACTERIA
DECOMPOSERS
FUNGUS
ANACONDA

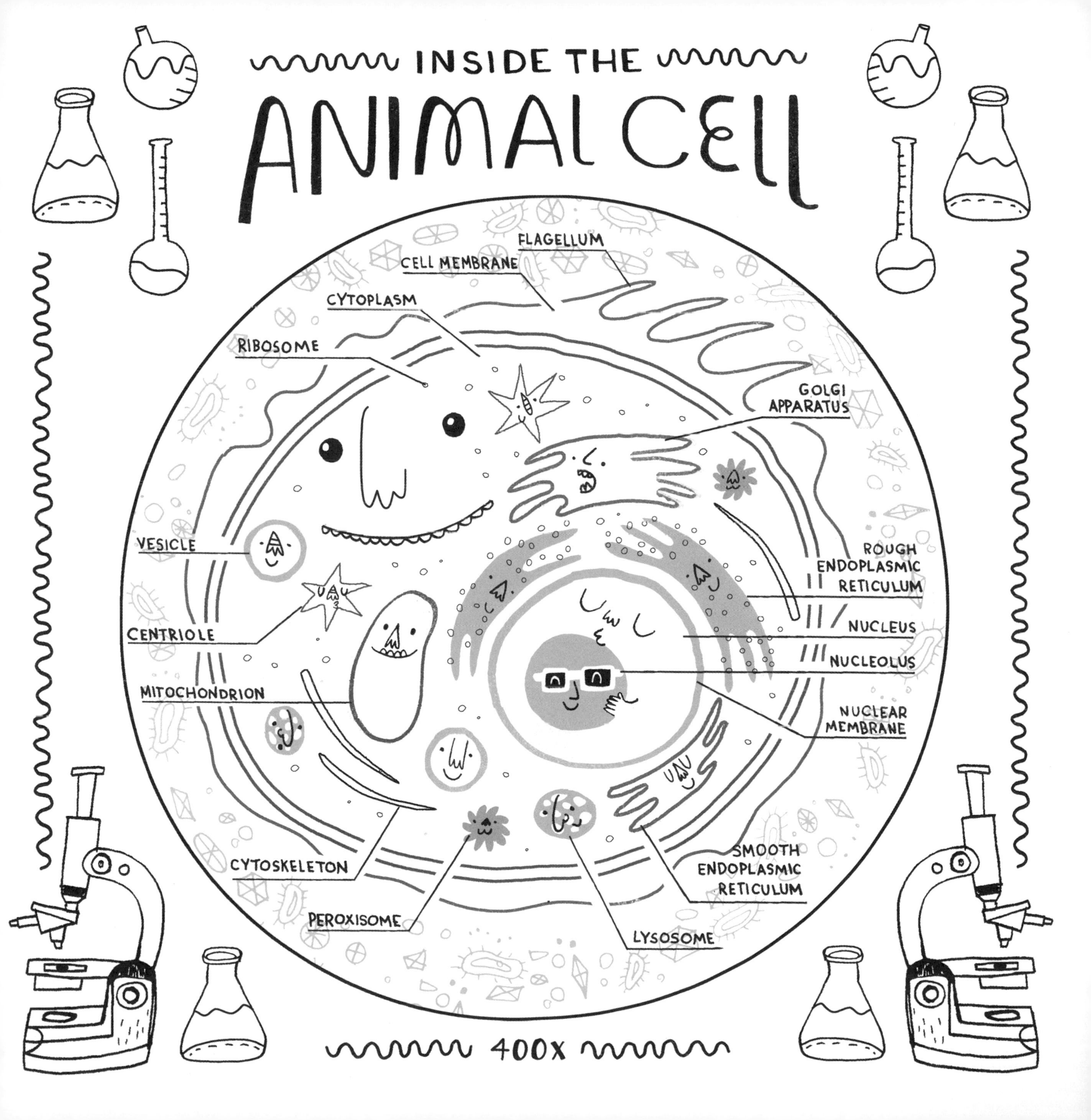
INSIDE THE
ANIMAL CELL
FLAGELLUM
CELL MEMBRANE
CYTOPLASM
RIBOSOME
GOLGI APPARATUS
VESICLE
ROUGH ENDOPLASMIC RETICULUM
CENTRIOLE
NUCLEUS
NUCLEOLUS
MITOCHONDRION
NUCLEAR MEMBRANE
CYTOSKELETON
SMOOTH ENDOPLASMIC RETICULUM
PEROXISOME
LYSOSOME
400X

ECOSYSTEM OF THE
ANDES
GLACIER
SCAVENGER
ANDEAN CONDOR
ALPINE GRASSLAND
3,000M–4,800M
ALPACA
PUYA
GIANT HUMMING-BIRD
GRASS
LLAMA
CHINCHILLA
CLOUD FOREST
800M–3,500M
YELLOW-TAILED WOOLLY MONKEY
MOTH
PALM LEAVES
FRUIT
ANDEAN TINAMOU
SPECTACLED BEAR
TROPICAL RAINFOREST
500M–1,500M
COMMON TREE BOA
GLASS FROG
DART FROG

ECOSYSTEM OF THE
ARCTIC CIRCLE
APEX PREDATOR
POLAR BEAR
EIDER DUCK
SCAVENGER
ARCTIC FOX
ICE ALGAE
ZOOPLANKTON
PRAWN
CONSUMERS
PHYTOPLANKTON
POLAR COD
RINGED SEAL
MELTING ARCTIC ICE
ICE EDGE ALGAE BLOOM
PRIMARY PRODUCER
WALRUS
NARWHAL
SQUID
PLANKTON
CLAMS
MUSSELS
BENTHOS
CRAB
DECOMPOSERS

"THE WILDERNESS
HOLDS ANSWERS
TO QUESTIONS
MAN HAS NOT
YET LEARNED
TO ASK."
- NANCY NEWHALL, EDITOR AND
PHOTOGRAPHY CRITIC

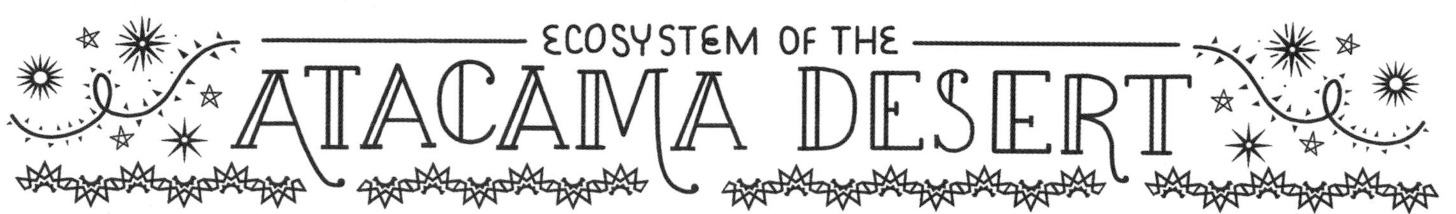
ECOSYSTEM OF THE
ATACAMA DESERT

VICUÑA
FLAMINGO
PAJA BRAVA GRASS
DIATOMS
SPECIAL ALGAE & PHYTOPLANKTON
SALT
SUPER ACIDIC SALTWATER LAKE
SCORPION
SOUTHERN VISCACHA
CULPEO FOX
COPIAPOA CACTUS
LEAF-EARED MOUSE

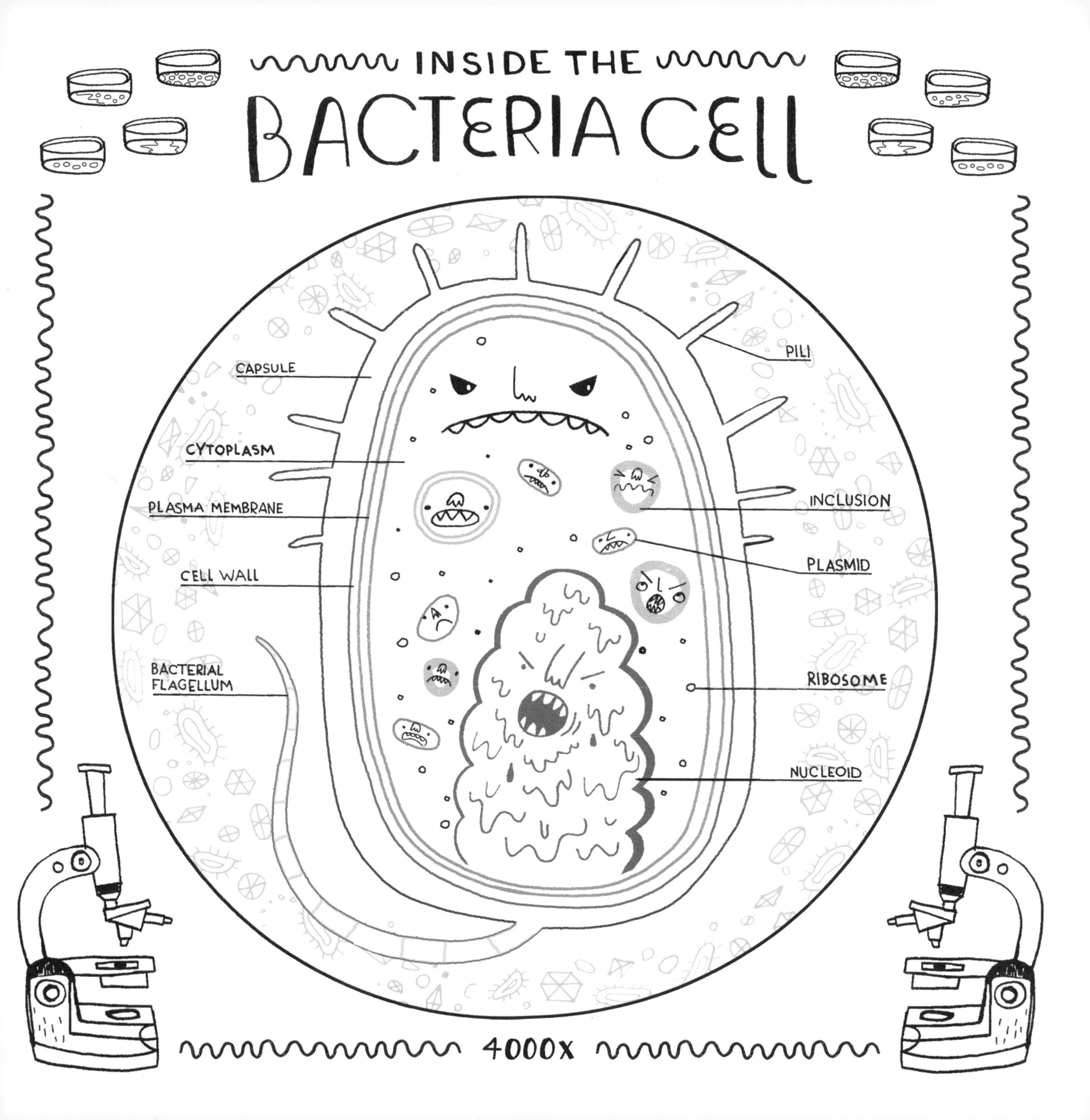
INSIDE THE
BACTERIA CELL
CAPSULE
CYTOPLASM
PLASMA MEMBRANE
CELL WALL
BACTERIAL FLAGELLUM
PILI
INCLUSION
PLASMID
RIBOSOME
NUCLEOID
4000x

ECOSYSTEM OF THE
CONGO RAINFOREST
CANOPY
LEAVES
GORILLA
CHIMPANZEE
UNDERSTORY
INSECTS
CONGO PEAFOWL
PALM
AFRICAN FOREST ELEPHANT
SHRUB LAYER
LEOPARD
OKAPI
DECOMPOSERS AND BACTERIA
BUTTRESS ROOT
FUNGUS
FOREST FLOOR

ECOSYSTEM OF THE
DEEP OCEAN
BLACK SEADEVIL ANGLERFISH
DUMBO OCTOPUS
BLUNTNOSE SIXGILL SHARK
VOLCANIC HOT SPRING VENTS
BARRELEYE
DEEP-SEA DRAGONFISH
VAMPIRE SQUID
BACTERIA AND ARCHAEA
TUBE WORMS
VENT MUSSELS
VENT OCTOPUS
EELPOUT
VENT CRAB
POMPEII WORM
FEATHER DUSTER WORMS

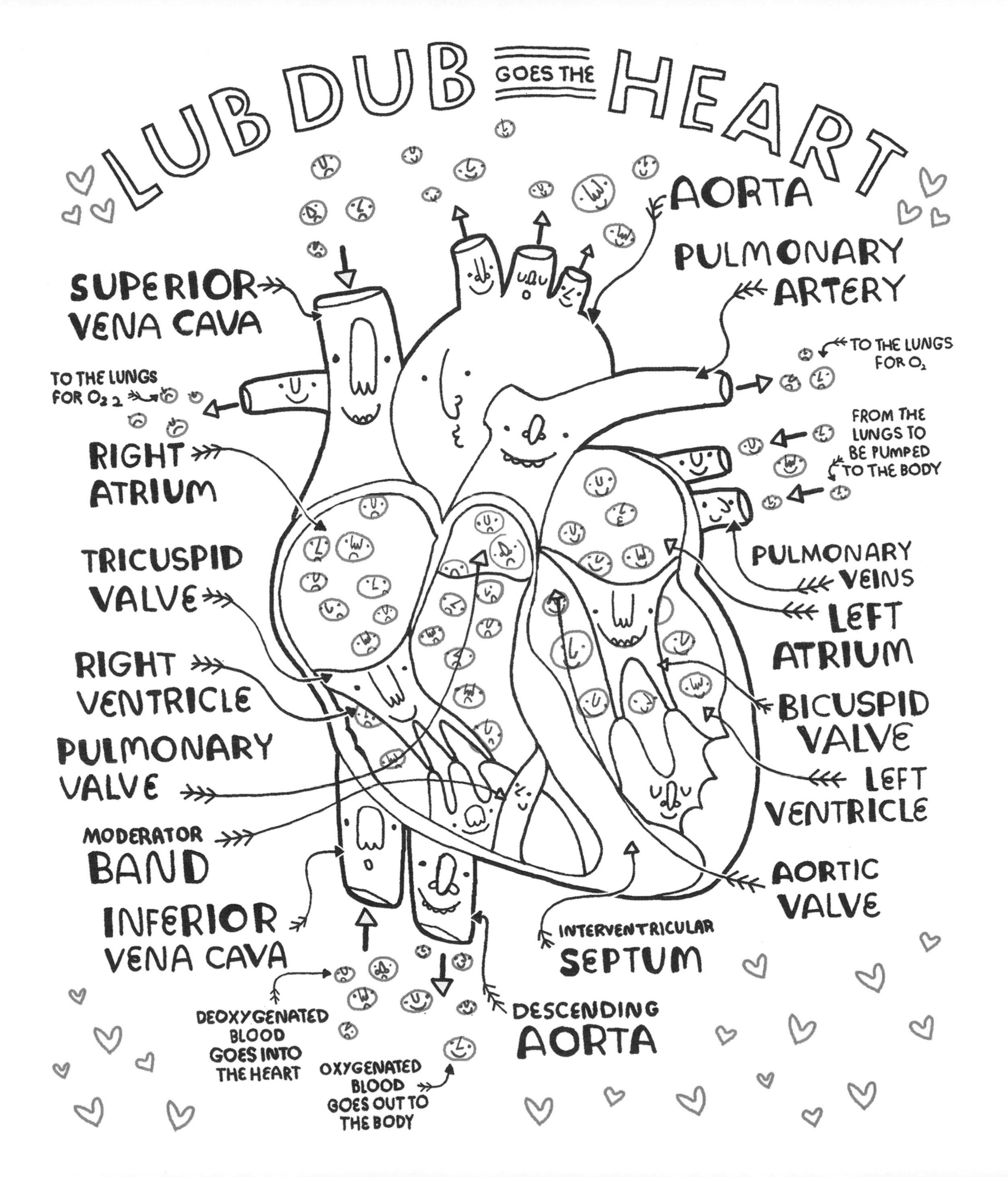
LUB DUB
GOES THE
HEART
AORTA
PULMONARY ARTERY
SUPERIOR VENA CAVA
TO THE LUNGS FOR O2
TO THE LUNGS FOR O2
FROM THE LUNGS TO BE PUMPED TO THE BODY
RIGHT ATRIUM
TRICUSPID VALVE
PULMONARY VEINS
LEFT ATRIUM
RIGHT VENTRICLE
BICUSPID VALVE
PULMONARY VALVE
LEFT VENTRICLE
MODERATOR BAND
AORTIC VALVE
INFERIOR VENA CAVA
INTERVENTRICULAR SEPTUM
DEOXYGENATED BLOOD GOES INTO THE HEART
DESCENDING AORTA
OXYGENATED BLOOD GOES OUT TO THE BODY

ECOSYSTEM
of A ROTTING LOG
CARPENTER MOTH
LICHEN
DOWNY WOODPECKER
WOOD
CARPENTER MOTH CATERPILLAR
TERMIT
OYSTER MUSHROOM
DRY ROT FUNGUS
CENTIPEDE
CARPENTER ANT
BARK BEETLE
SAP
APHIDS
HONEY-DEW
BACTERIA

A DESERT ECOSYSTEM
APEX PREDATOR
RED-TAILED HAWK
MOTH
ELF OWL
PRICKLY PEAR CACTUS
SAGUARO CACTUS
GILA WOOD-PECKER
KANGAROO RAT
KIT FOX
BRITTLEBUSH
DIAMONDBACK RATTLESNAKE
PRODUCERS
SECONDARY CONSUMERS
ARMADILLO
ANTS
MILLIPEDES
DUNG BEETLE
SCORPION
GRASSHOPPER
COLLARED LIZARD
DÉCOMPOSERS

"YOU CAN'T CONSERVE WHAT YOU HAVEN'T GOT."
- MARJORY STONEMAN DOUGLAS,
CONSERVATIONIST AND AUTHOR

ECOSYSTEM OF THE
GREAT PLAINS
MIGRATORY BIRD
SANDHILL CRANE
BISON
AWOOoooo
COYOTE
PRONGHORN
WESTERN MEADOWLARK
BLACK-TAILED PRAIRIE DOG
BURROWING OWL
APHID
BUFFALO GRASS
BIG BLUESTEM GRASS
GRASS
WOLF SPIDER
GRASSHOPPER
CONEFLOWER
MICROBES
FUNGUS
DECOMPOSERS
WORM
STOLEN PRAIRIE DOG NEST
OWL EGGS

ECOSYSTEM OF THE
GREAT BARRIER REEF
APEX PREDATOR
HAMMERHEAD SHARK
SOFT CORAL
STINGRAY
GREEN SEA TURTLE
SEAHORSE
BUTTERFLYFISH
RED-SPOTTED CORAL CRAB
CLOWNFISH
CORAL TROUT
ZOOPLANKTON
PARROTFISH
SEA ANEMONE
ALGAE
SURGEONFISH
CORAL POLYP
TENTACLE
ZOOXANTHELLAE
MOUTH
BASAL PLATE
SKELETON
BRAIN CORAL
OCTOPUS
CROWN-OF-THORNS STARFISH
SEA URCHIN
HARD CORAL

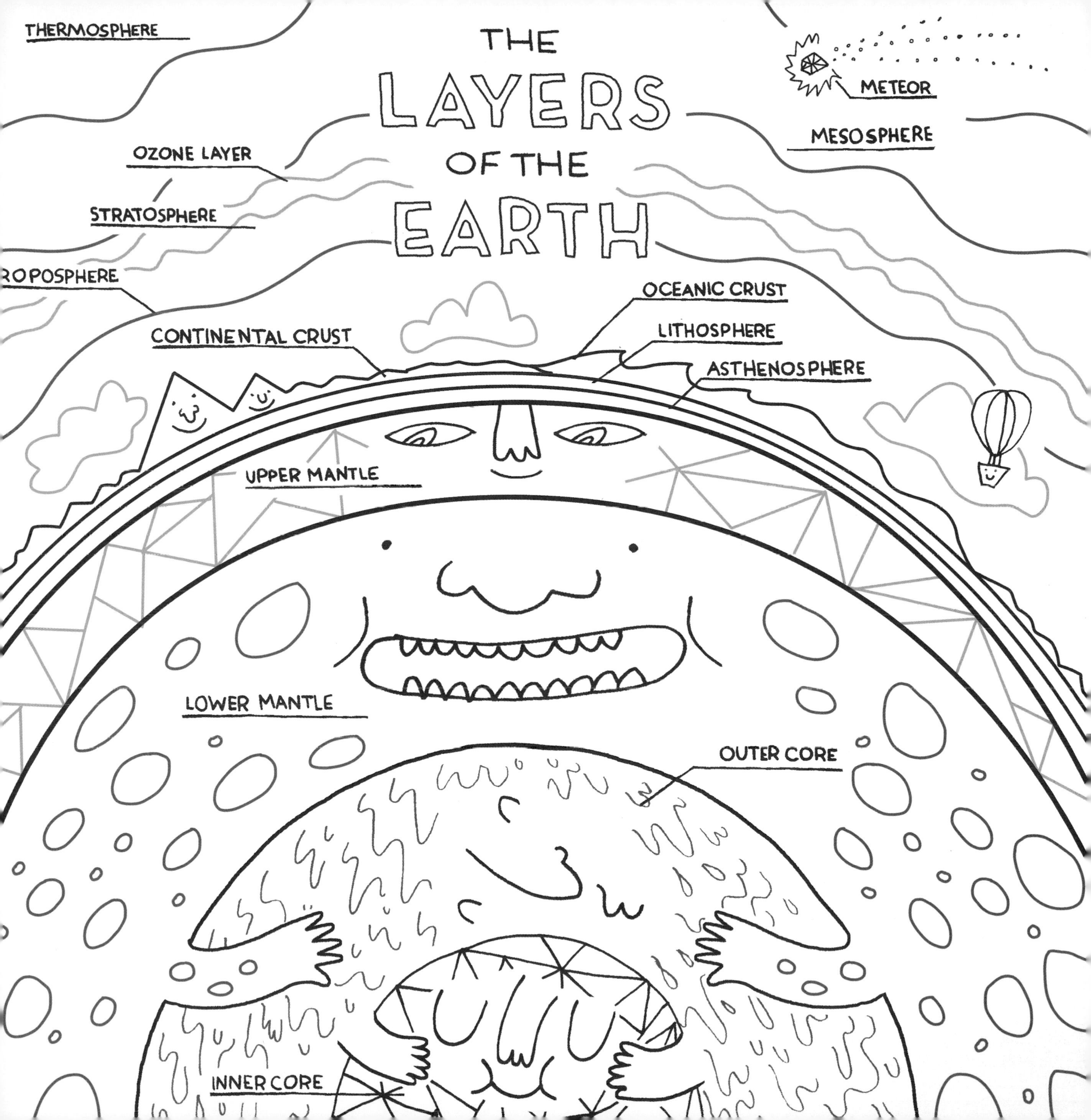
THE LAYERS OF THE EARTH
THERMOSPHERE
METEOR
MESOSPHERE
OZONE LAYER
STRATOSPHERE
ROPOSPHERE
OCEANIC CRUST
CONTINENTAL CRUST
LITHOSPHERE
ASTHENOSPHERE
UPPER MANTLE
LOWER MANTLE
OUTER CORE
INNER CORE

"THOSE WHO
DWELL AMONG
THE BEAUTIES
AND MYSTERIES
OF THE EARTH
ARE NEVER
ALONE OR
WEARY OF LIFE."
—RACHEL CARSON, MARINE BIOLOGIST,
CONSERVATIONIST, AND AUTHOR

ECOSYSTEM OF THE
MANGROVE SWAMP
RACCOON
IGUANA
ROSEATE SPOONBILL
GREAT EGRET
PRODUCER
RED MANGROVE
APEX PREDATOR
AMERICAN ALLIGATOR
AMERICAN CROCODILE
FISH NURSERY
LONGNOSE GAR
SEA TURTLE
DEAD LEAVES
MANGROVE SNAPPER
FISH EGGS
DEAD LEAVES
PRAWN
CRAYFISH
SEA GRASS
CRAB
MOLLUSKS
AMPHIPOD
POLYCHAETE WORM

ECOSYSTEM OF THE

MOORLANDS OF THE BRITISH ISLES

SPHAGNUM (PEAT MOSS)
DEAD VEGETATION
GROUND-WATER
PRESSURE AND HEAT
PEAT FORMS
RED DEER
MERLIN
BOG
EGGS
BADGER
MOUNTAIN HARE
GRASS
SKYLARK
PEAT MOSS
RED GROUSE
WORM
PRODUCER HEATHER
EMPEROR MOTH
DRAGONFLY
BUMBLEBEE
VERTIGO SNAIL
COTTON-GRASS
SNIPE

"ONLY
WHEN OUR
CLEVER BRAIN
AND OUR
HUMAN HEART
WORK TOGETHER
IN HARMONY
CAN WE ACHIEVE
OUR FULL
POTENTIAL."
–JANE GOODALL, PRIMATOLOGIST,
ETHNOLOGIST, AND ANTHROPOLOGIST

ECOSYSTEM OF THE
OPEN OCEAN
HUMPBACK WHALE
PHYTOPLANKTON
PRODUCERS
SUNLIGHT ZONE
KRILL
HERRING
ZOOPLANKTON
MACKEREL
BOTTLENOSE DOLPHIN
BLUEFIN TUNA
COPEPODS
ATLANTIC BLUE MARLIN
GREAT WHITE SHARK
APEX PREDATOR
SQUID
TWILIGHT ZONE
EEL
JELLYFISH
BRISTLEMOUTH FISH

ECOSYSTEM OF THE
SAHARA DESERT
DATE PALM
OASIS
SAND DUNE
DESERT CROCODILE
CAMEL
DUNG BEETLE
CAMEL POOP
ROSE OF JERICHO
DEATH STALKER SCORPION
LOVEGRASS
MONITOR LIZARD
FENNEC FOX
JERBOA
SAHARAN SILVER ANT
SCAVENGERS
SAHARAN HORNED VIPER

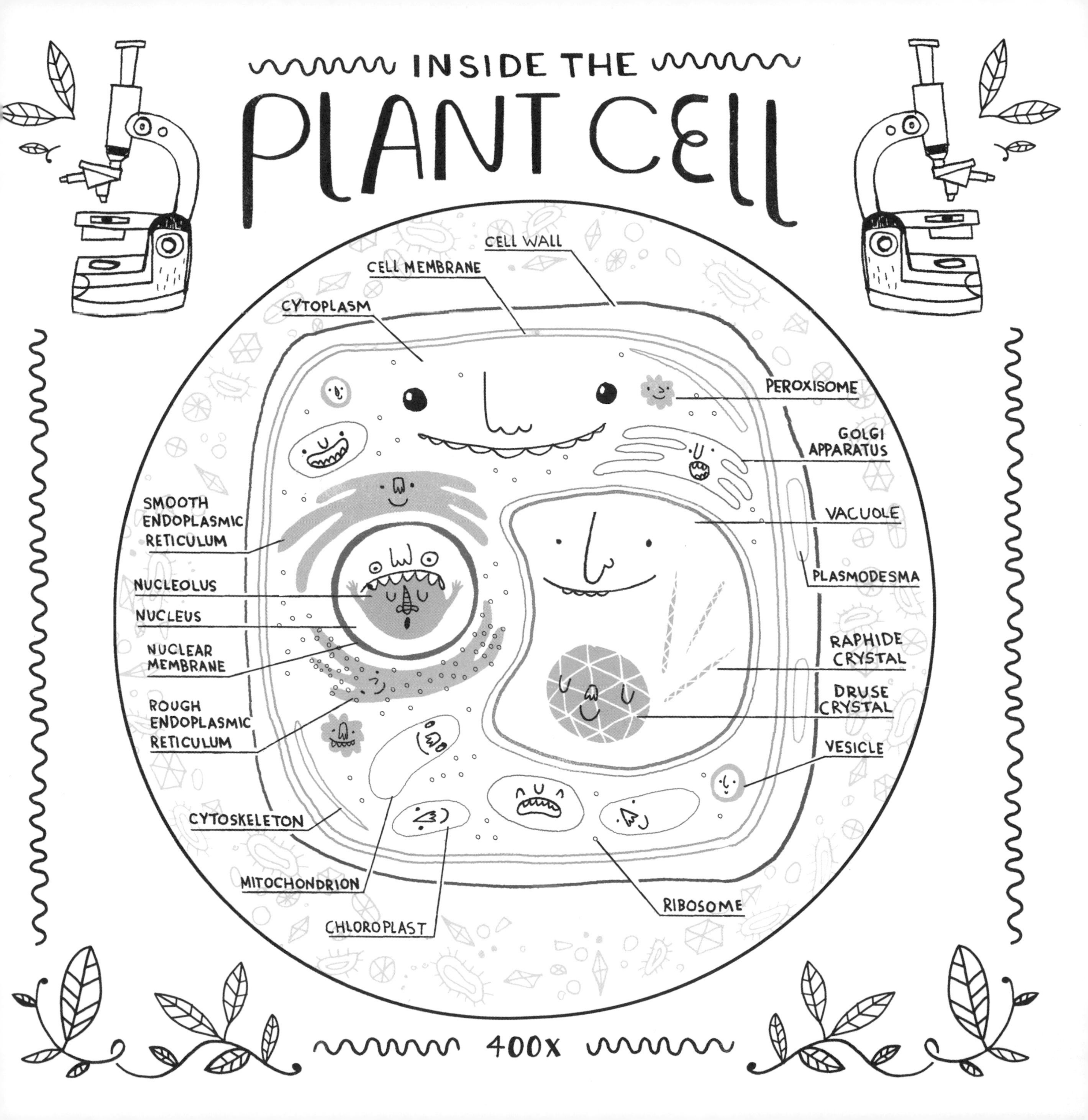
INSIDE THE
PLANT CELL
CELL WALL
CELL MEMBRANE
CYTOPLASM
PEROXISOME
GOLGI APPARATUS
SMOOTH ENDOPLASMIC RETICULUM
VACUOLE
PLASMODESMA
NUCLEOLUS
NUCLEUS
NUCLEAR MEMBRANE
RAPHIDE CRYSTAL
DRUSE CRYSTAL
ROUGH ENDOPLASMIC RETICULUM
VESICLE
CYTOSKELETON
MITOCHONDRION
RIBOSOME
CHLOROPLAST
400X

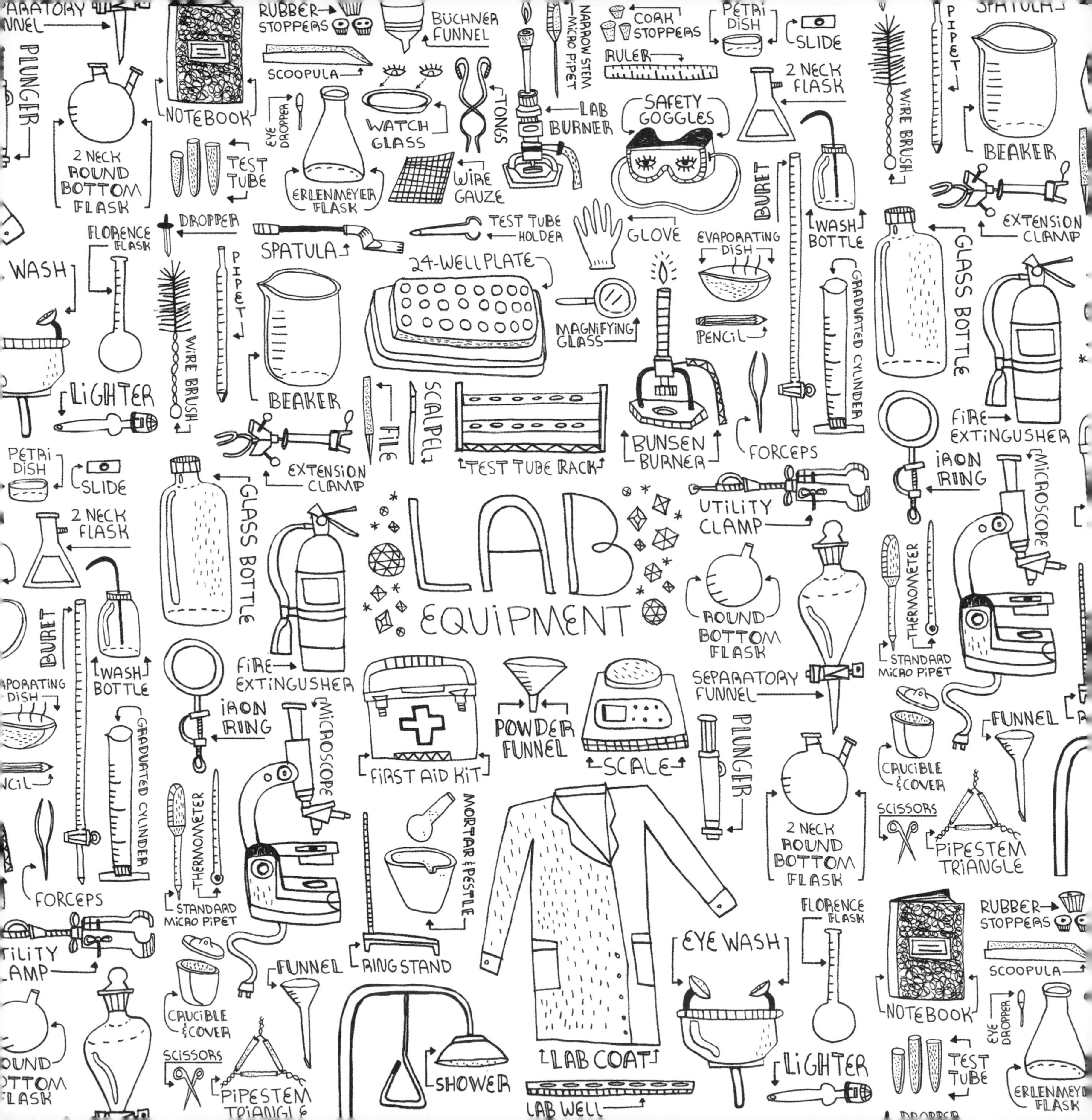

LAB
EQUIPMENT
PLUNGER
NOTEBOOK
RUBBER STOPPERS
BÜCHNER FUNNEL
SCOOPULA
NARROW STEM MICRO PIPET
CORK STOPPERS
PETRI DISH
SLIDE
RULER
PIPET
BEAKER
2 NECK ROUND BOTTOM FLASK
EYE DROPPER
WATCH GLASS
TONGS
LAB BURNER
SAFETY GOGGLES
2 NECK FLASK
WIRE BRUSH
TEST TUBE
ERLENMEYER FLASK
WIRE GAUZE
BURET
WASH BOTTLE
EXTENSION CLAMP
DROPPER
FLORENCE FLASK
SPATULA
TEST TUBE HOLDER
GLOVE
EVAPORATING DISH
GLASS BOTTLE
24-WELL PLATE
MAGNIFYING GLASS
PENCIL
GRADUATED CYLINDER
LIGHTER
FILE
SCALPEL
TEST TUBE RACK
BUNSEN BURNER
FORCEPS
FIRE EXTINGUSHER
IRON RING
MICROSCOPE
UTILITY CLAMP
ROUND BOTTOM FLASK
THERMOMETER
STANDARD MICRO PIPET
FIRST AID KIT
POWDER FUNNEL
SCALE
SEPARATORY FUNNEL
FUNNEL
CRUCIBLE & COVER
MORTAR & PESTLE
SCISSORS
PIPESTEM TRIANGLE
EYE WASH
RING STAND
SHOWER
LAB COAT
LAB WELL

ECOSYSTEM OF A POND

ECOSYSTEM OF THE REDWOOD FOREST

ECOSYSTEMS OF
RIVERS
HEADWATERS
(A RIVER'S SOURCE)
OSPREY
MOOSE
FLOODPLAIN
TRIBUTARY
RIVER
CHANNEL
AQUATIC
PLANTS
RIVER OTTER
PLANKTON
AND ALGAE
RIVER BANK
WOOD
DUCK
FALLEN
ACORNS AND
LEAVES
ARROWHEAD
BLUEGILL
ZOOPLANKTON
SMALL
FISH
CRAYFISH
DRAGONFLY
FROG
SMALL
FISH
MAY-
FLIES
RAINBOW
TROUT
DELTA
OCEAN
FLATHEAD CATFISH
DECOMPOSERS
GROUND WATER
ESTUARIES

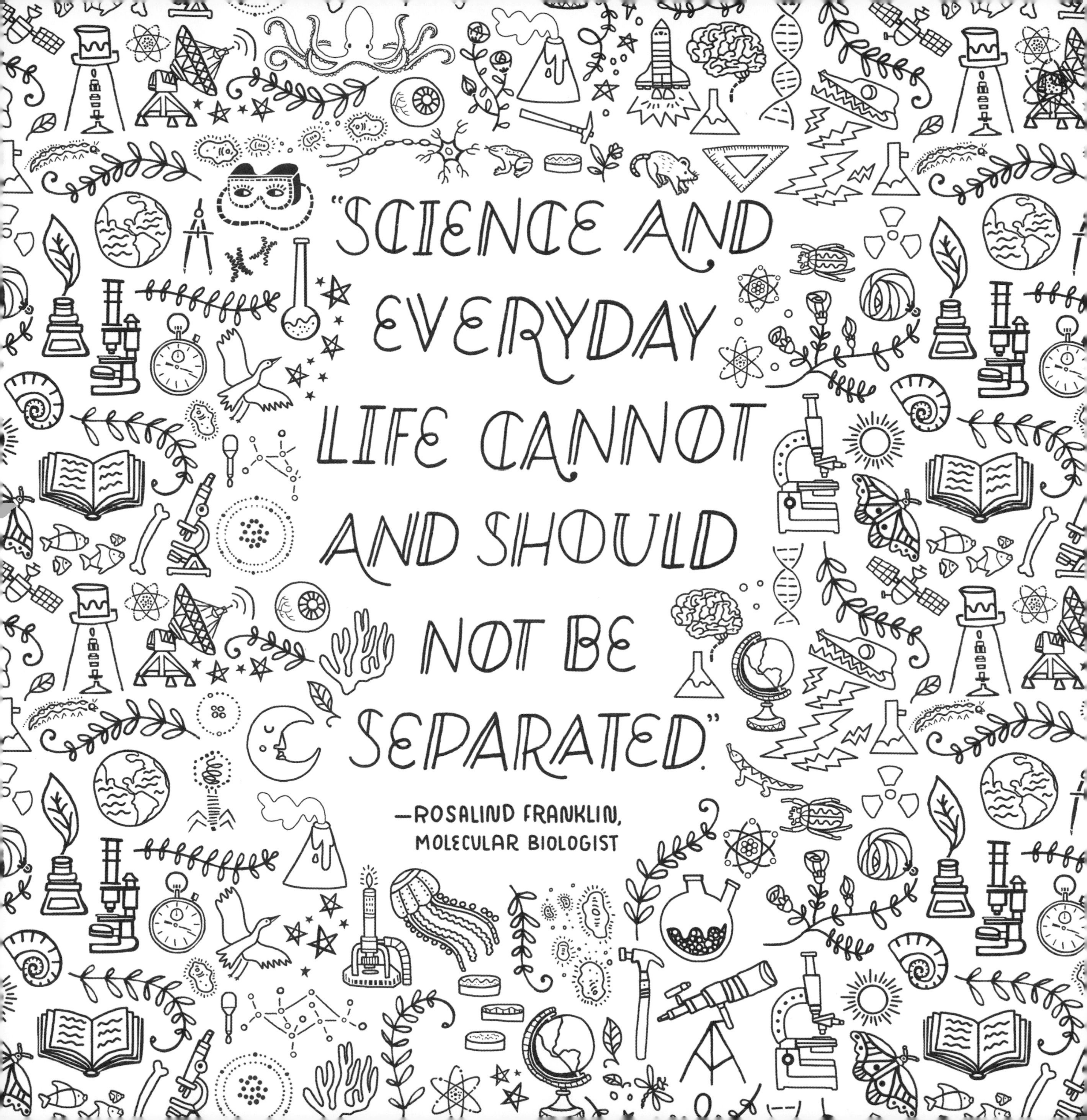
"SCIENCE AND
EVERYDAY
LIFE CANNOT
AND SHOULD
NOT BE
SEPARATED."
—ROSALIND FRANKLIN,
MOLECULAR BIOLOGIST

ECOSYSTEM OF THE
SIBERIAN TAIGA
GREAT GRAY OWL
PRODUCERS
FIR TREE
REINDEER
SAXIFRAGA (STONE-BREAKER)
SPRUCE TREE
LYNX
LICHEN
ERMINE
SIBERIAN BROWN BEAR
TAIGA VOLE
MOSS
ARCTIC HARE
GRASS

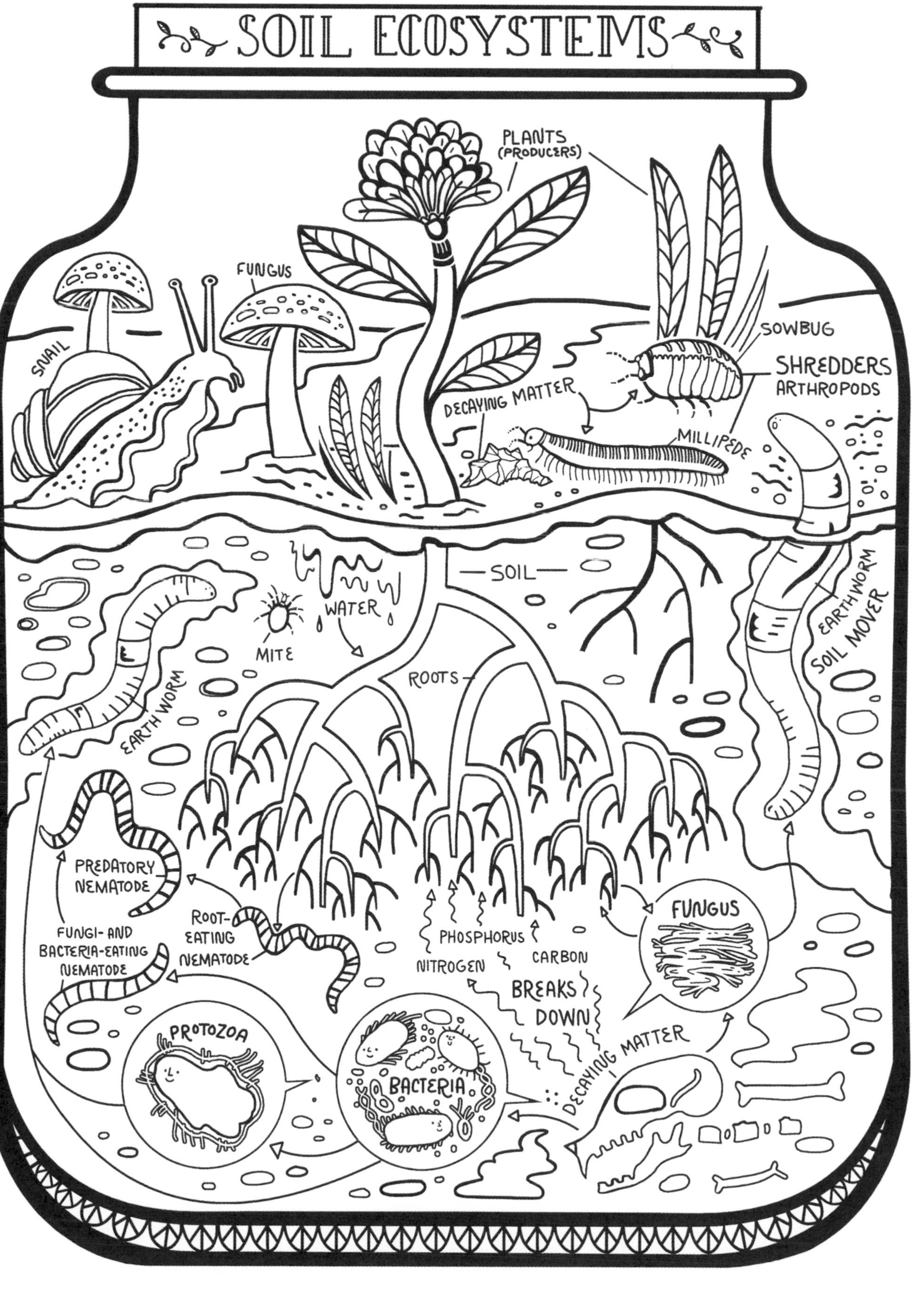

SOIL ECOSYSTEMS

PLANTS (PRODUCERS)
FUNGUS
SNAIL
SOWBUG
SHREDDERS ARTHROPODS
DECAYING MATTER
MILLIPEDE
SOIL
WATER
MITE
EARTHWORM SOIL MOVER
ROOTS
EARTH WORM
PREDATORY NEMATODE
ROOT-EATING NEMATODE
FUNGI- AND BACTERIA-EATING NEMATODE
PHOSPHORUS
NITROGEN
CARBON
BREAKS DOWN
FUNGUS
PROTOZOA
BACTERIA
DECAYING MATTER

PLANTS

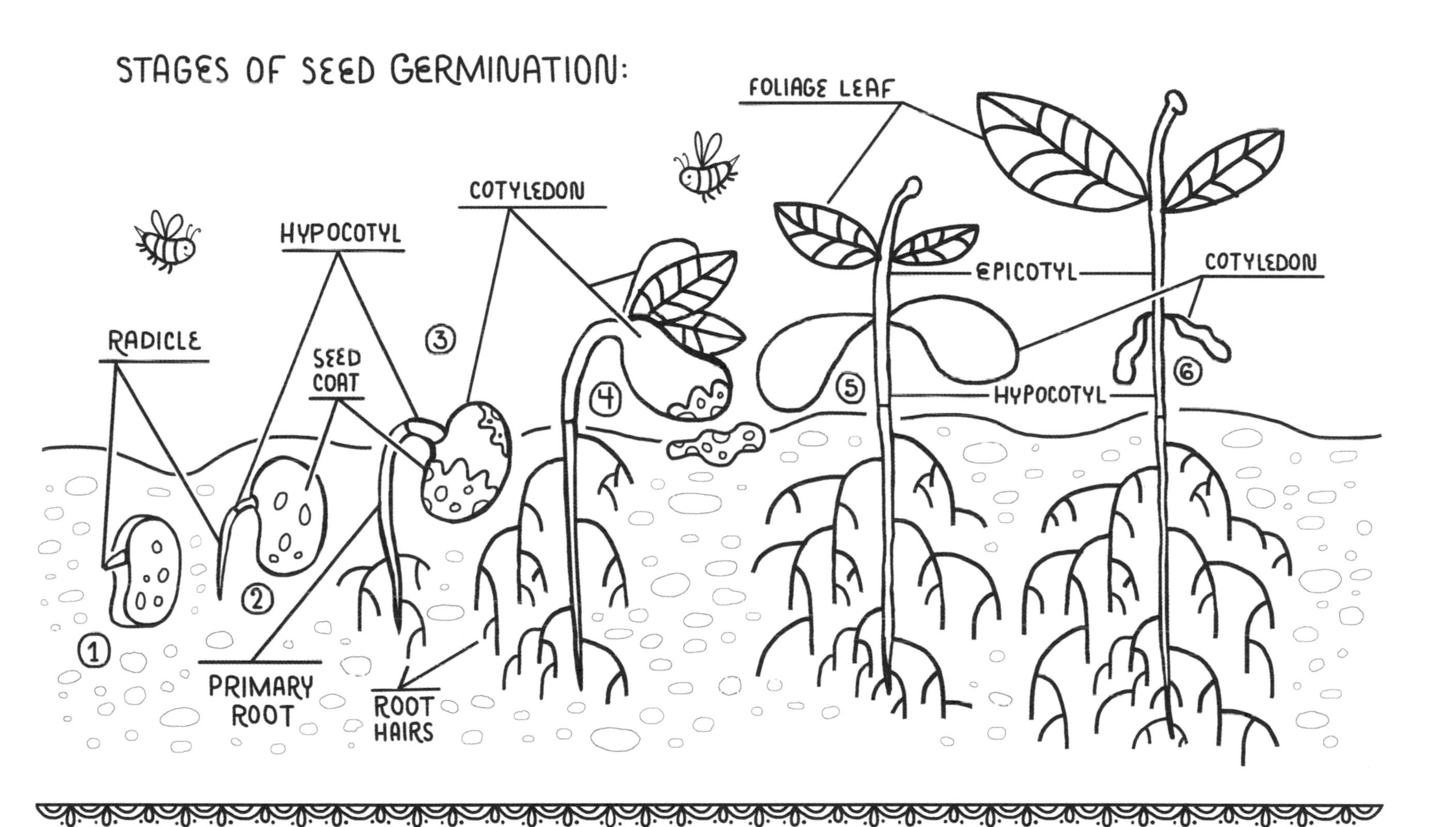
STAGES OF SEED GERMINATION:
FOLIAGE LEAF
COTYLEDON
HYPOCOTYL
EPICOTYL
COTYLEDON
RADICLE
SEED COAT
3
4
5
6
HYPOCOTYL
2
1
PRIMARY ROOT
ROOT HAIRS

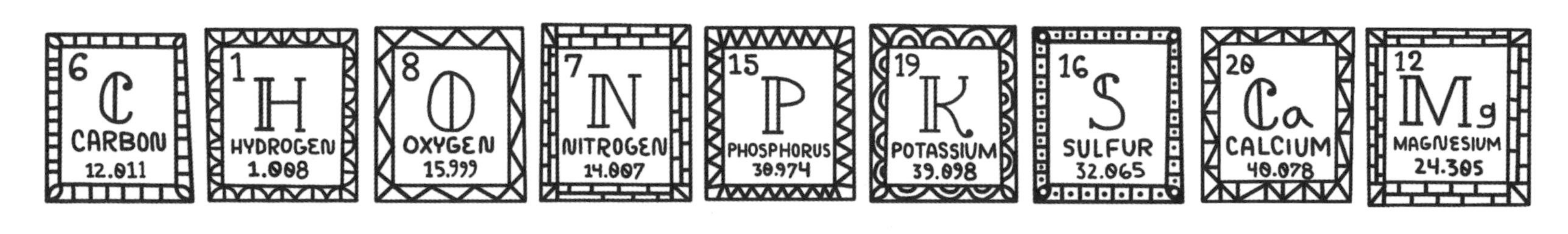
MACRONUTRIENTS PLANTS NEED:
6 C CARBON 12.011
1 H HYDROGEN 1.008
8 O OXYGEN 15.999
7 N NITROGEN 14.007
15 P PHOSPHORUS 30.974
19 K POTASSIUM 39.098
16 S SULFUR 32.065
20 Ca CALCIUM 40.078
12 Mg MAGNESIUM 24.305

"YOU CAN CHANGE THE WORLD, FOR YOU ARE MADE OF STAR STUFF, AND YOU ARE CONNECTED TO THE UNIVERSE."
—VERA RUBIN, ASTRONOMER

ECOSYSTEM OF THE TASMANIAN
RAINFOREST
PRODUCER
MOUNTAIN ASH TREE
CLIMBING HEATH
MACLEAY'S SWALLOWTAIL BUTTERFLY
MOSS
HARD WATER FERN
WORMS AND ANTS
ECHIDNA
TASMANIAN ROSELLA
TASMANIAN LAUREL
SPOTTED-TAIL QUOLL
PLATYPUS
ANNELID WORM
SHRIMP
VELVET WORM
SLIME
SCAVENGER
TASMANIAN DEVIL
WOMBAT
CRICKET

MICROSCOPIC ECOSYSTEMS OF A WATER DROP
DIATOMS
PHYTOPLANKTON (PRODUCERS)
ALGAE
CYANOBACTERIA
JUVENILE SQUID
COPEPODS
RADIOLARIA
ZOOPLANKTON
ARROW WORM
AMPHIPODA
KRILL

THE BIG WORLD
WE LIVE IN IS
SMALLER THAN
YOU THINK...

PROTECTING OUR PLANET

Published in the United States by Clarkson Potter/Publishers, an imprint of Random House, a division of Penguin Random House LLC, New York.

clarksonpotter.com

Selected artwork is based on illustrations found in *I Love Science* (Ten Speed Press, 2017), copyright © 2017 by Rachel Ignotofsky; *The Wondrous Workings of Planet Earth* (Ten Speed Press, 2018), copyright © 2018 by Rachel Ignotofsky; and *I Love the Earth* (Clarkson Potter, 2020), copyright © 2020 by Rachel Ignotofsky.

CLARKSON POTTER is a trademark and POTTER with colophon is a registered trademark of Penguin Random House LLC.

ISBN 978-0-593-23314-6

Printed in China

Illustrations and cover design by Rachel Ignotofsky

10 9 8 7 6 5

First Edition